SENIORS

KATHRYN ANN SAID

"She has a wonderful gift of making friends"

Student Council 1, 2, 3, 4, President 4; Center Staff 1; Red Rectangle 2; Y-Teens 3, 4; Pharmacy Club 4; Girls' Athletic Association 1, 2; Boosters Club 1, 2, 3, 4; 4-H Club 1, 2, 3; Economics Club 3, Vice-President 3; Home Vice-President 3.

NEIL A. ARMSTRONG
"He thinks, he acts, 'tis done."

Band 2, 3, 4, Vice-President 4; Orchestra 3; Glee Club 2; Student Council 3, 4, Vice-President 4; Retrospect Staff; Junior Hi-Y 2; Senior Hi-Y 3, 4; Boosters Club 2, 3, 4; Junior Class Play; Home Room President 3; Boys' State 3; Transferred from Upper Sandusky High School 1.

Neil A. Armstrong

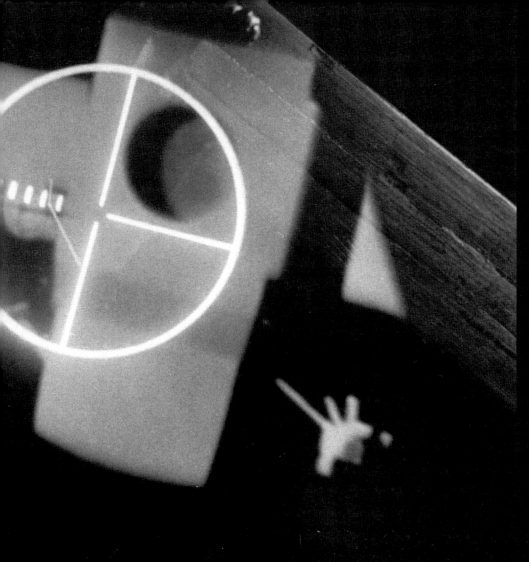

LIFE BOOKS
Managing Editor Robert Sullivan
Director of Photography Barbara Baker Burrows
Creative Director Mimi Park
Deputy Picture Editor Christina Lieberman
Copy Chief Barbara Gogan
Copy Editors Don Armstrong, Parlan McGaw
Writer-Reporters Marilyn Fu, Amy Lennard Goehner
Photo Associate Sarah Cates
Editorial Associate Courtney Mifsud
Consulting Picture Editors
Mimi Murphy (Rome), Tala Skari (Paris)

Editorial Director Stephen Koepp
Editorial Operations Director Michael Q. Bullerdick

EDITORIAL OPERATIONS
Richard K. Prue (Director), Brian Fellows (Manager),
Richard Shaffer (Production), Keith Aurelio,
Charlotte Coco, Kevin Hart, Mert Kerimoglu,
Rosalie Khan, Patricia Koh, Marco Lau, Brian Mai,
Po Fung Ng, Rudi Papiri, Robert Pizaro, Barry Pribula,
Clara Renauro, Katy Saunders, Hia Tan, Vaune Trachtman

TIME HOME ENTERTAINMENT
Publisher Jim Childs
Vice President, Business Development & Strategy
Steven Sandonato
Executive Director, Marketing Services Carol Pittard
Executive Director, Retail & Special Sales Tom Mifsud
Executive Publishing Director Joy Butts
Director, Bookazine Development & Marketing
Laura Adam
Finance Director Glenn Buonocore
Associate Publishing Director Megan Pearlman
Assistant General Counsel Helen Wan
Assistant Director, Special Sales Ilene Schreider
Book Production Manager Suzanne Janso
Design & Prepress Manager Anne-Michelle Gallero
Brand Manager Roshni Patel
Associate Prepress Manager Alex Voznesenskiy
Assistant Brand Manager Stephanie Braga

Special thanks: Christine Austin, Katherine Barnet,
Jeremy Biloon, Susan Chodakiewicz, Rose Cirrincione,
Lauren Hall Clark, Jacqueline Fitzgerald, Christine Font,
Jenna Goldberg, Hillary Hirsch, David Kahn,
Amy Mangus, Robert Marasco, Kimberly Marshall,
Amy Migliaccio, Nina Mistry, Dave Rozzelle,
Ricardo Santiago, Adriana Tierno, Vanessa Wu

Copyright © 2012
Time Home Entertainment Inc.

Published by LIFE BOOKS, an imprint of
Time Home Entertainment Inc.,
135 West 50th Street, New York, New York 10020

ISBN 10: 1-61893-073-7
ISBN 13: 978-1-61893-073-6
Library of Congress Control Number: 2012948057

Vol. 12, No. 21 • September 7, 2012

"LIFE" is a registered trademark of Time Inc.

We welcome your comments and suggestions about
LIFE Books. Please write to us at: LIFE Books,
Attention: Book Editors, PO Box 11016,
Des Moines, IA 50336-1016

If you would like to order any of our hardcover Collector's
Edition books, please call us at 1-800-327-6388.
(Monday through Friday, 7:00 a.m.–8:00 p.m.
or Saturday, 7:00 a.m.–6:00 p.m. Central Time).

TABLE OF CONTENTS

FRONT COVER: THE ASTRONAUT IN 1969, PHOTOGRAPH BY NASA/ZUMA
BACK COVER: APOLLO 11 LAUNCH, JULY 16, 1969, PHOTOGRAPH BY
RALPH CRANE
ENDPAPERS: BOOT PRINT ON THE MOON, JULY 20, 1969, PHOTOGRAPH BY
BUZZ ALDRIN/CORBIS
PAGE 1: EARTH AND MOON AS SEEN FROM APOLLO 11, PHOTOGRAPH BY NASA/
PHOTO RESEARCHERS/GETTY
PAGES 2–3: NEIL ARMSTRONG'S BLUME HIGH SCHOOL YEARBOOK ENTRY,
PHOTOGRAPH BY DAVID HOWELLS/CORBIS
THESE PAGES: IN TRAINING, 1964, PHOTOGRAPH BY BETTMANN/CORBIS

Foreword

By Jim Lovell

How do you write a foreword about an icon?

It would be difficult to identify another American who achieved as much
in his lifetime as Neil Armstrong, or one who ascended to the heights
of global fame and recognition that resulted from his remarkable successes.
In spite of it all, he remained a humble man and used his fame not
for personal gain but to contribute to the legacy of America in the world.

His contributions to his country were many—a Korean War naval aviator,
a government test pilot, a NASA astronaut, a college professor, a successful
businessman and an American who didn't hesitate to express his views.

He will be missed.

THOMAS M. LEE/MORALE ENTERTAINMENT FOUNDATION

RALPH MORSE

BOUND BY GLORY At top left on the opposite page is Neil Armstrong. At bottom left is James Lovell Jr., the late Armstrong's friend as well as the writer of the foreword on this page. The year is 1963, and this is the new crop of NASA astronauts who will be asked to follow in the space boots of the legendary Mercury 7. For Armstrong, the adventures on *Gemini 8* and *Apollo 11,* the latter of which will take him to his destiny as the first man on the moon, are well ahead, not even yet a dream. Also in the unseeable future are Jim Lovell's two Gemini and two Apollo missions, the last of which will be as commander of *Apollo 13,* which will be stricken en route to the moon, requiring Lovell's steady hand to bring the crew home. The others in this group of hopefuls in '63 are, to the right of Lovell, James McDivitt and Charles Conrad Jr.; in the second row, Elliot See Jr. (who will be killed in training during Project Gemini) and Thomas Stafford; in the third row Edward White II and John Young; and alongside Armstrong, Frank Borman. In the photograph on this page, Lovell and Armstrong are together again in 2010, in Ramstein Air Base in Germany, during a tour to boost morale among Americans serving abroad.

Introduction
The Man Who Leapt

TODAY, IT IS EASILY IMAGINED that we might step foot on Mars one day. It's almost as if we've already done so, with the first inquisitive, keen-sighted Rover and his (or her) intrepid next-generation offspring having been deployed so successfully.

Mightn't we soon float perpetually in the heavens, going about our day-to-day business like the folks on *Star Trek*? Or haven't we, in fact, been doing that for more than a decade with the International Space Station? Sure, it looks to be more cramped than the *Enterprise*—a jalopy alongside Captain Kirk's stretch limo— but it's a start. And it has the added advantage of being real.

Human aspiration has been largely stripped of the notion of impossibility. We have turned so many of yesteryear's unthinkable adventures and science fictions into historical accomplishments that we have come to feel that even the sky is no limit. Certainly in the future we will go amidst the stars, where now only the Hubble telescope can take us. Certainly the athlete of tomorrow, genetically engineered, will break the three-minute mile.

But once, and not too very long ago, the world—make that the universe—was filled with impossibilities. A journey to the bottom of the sea and a trip to the moon were the imaginings of fantasists like Jules Verne, as improbable as Tolkien's hobbits or Rowling's Hogwarts.

Little boys who grew up on Verne began to wonder, and grew to become America's and Russia's early rocketeers. Kids raised on the Tom Swift books sensed the moon growing closer. Meantime, impossibilities were ticked off the list—Peary reached the North Pole, Hillary and Tenzing conquered Everest, Bannister broke the four-minute mile, man descended to the deepest and inkiest depths of the Marianas Trench—and the whole idea of *impossible* was under assault.

Neil Armstrong, who was born into this constantly changing world in Ohio in 1930, was much like other boys, as we have been reminded in recent days. He was shy,

dutiful, likable and calm—preternaturally calm under pressure, as it would turn out. His colleague pilots in the space program would come to consider him the coolest of them all, a kinsman of the Chuck Yeager character as portrayed by Sam Shepard in the film version of *The Right Stuff*. A year before he walked on the moon, Armstrong was nearly killed when he ejected from a landing module with barely a second and a half to spare. When asked if the incident—one of his several near-death experiences—had really happened like that, he said only, "Yeah."

He began his NASA career at the Lewis Research Center, later renamed Glenn Research Center for an earlier midwestern space cowboy, and did well in the Gemini and Apollo programs. He had been largely unnoticed by the public before he was

BORN TO WALK
Above, in Ohio in 1933, the future first man to walk on the moon is taking only small steps—and not yet for mankind— at age three. Opposite: On July 16, 1969, at Cape Kennedy, Florida, the *Apollo 11* commander is striding purposefully forth, leading his fellow crew members to their destiny—a trip that, beginning today, will lead to the lunar surface, and thus into the annals of history.

anointed to be the one: the first man on the moon. With the Mercury 7 astronauts of the late '50s, everyone had known all about this gang of heroes—Shepard, Glenn, Grissom, Cooper, et al.—through the pages of LIFE. The reading public knew all about their wives, even. Now, in 1969, we had to catch up with Neil Armstrong, and we did so, and we do so again in the pages of LIFE. He was someone worth knowing back then, and he is certainly someone worth remembering now.

He was the right man for the job. That quote of his as he was in the act of placing his booted foot on the lunar surface was confirmation: He understood what was happening, and he had a philosopher's view and something of a poet's soul to go along with his crack-pilot talents. He brought us into the moment. He not only allowed but urged us all— Russians, Americans and all the peoples of the earth—to take the step with him, to make the leap. He took what was already conceived as a transcendent moment in a tumultuous time and, unbelievably, he elevated it. He wasn't just an astronaut up there, and this wasn't just about science, or the cold war, or any other one thing. It was about possibility, and . . .

E.B. White wrote at the time, "One-sixth gravity must be a lot of fun, and when Armstrong and Aldrin went into their bouncy little dance, like two happy children, it was a moment not only of triumph but of gaiety."

Yes, and of joy.

Triumph, gaiety, joy, amazement . . . and possibility.

The story line goes that Armstrong retreated from view after his famous adventure, but he didn't. He spoke often enough, to whom he wanted to, about the moon walk and about his life, as we hope will be made clear in our book. Looked at one way—born in Ohio, successful career, three kids, 10 grandkids, quiet retirement, died in Ohio at age 82—it looks like a nice, to-be-wished-for life's journey.

But what a journey it truly was, and he was generous enough to take us along.

Eagle Scout

IN PICTURES that you have already seen in this book, you have been clued in to much that you need to know about Neil Armstrong. First, there was the image of the earth and moon taken during the historic 1969 Apollo 11 mission, during which Armstrong became the first man ever to set foot on our planet's satellite and nearest neighbor. "The first man on the moon": This was how he would always be defined by others, though he never let the achievement define him. Because, as the very next illustration tells us, he spent his lifetime achieving, and then moving resolutely on to the next accomplishment. Consider the testimonial in that yearbook entry on pages two and three of our book, which was bestowed by his 1947 senior year classmates at Blume High School: "He thinks, he acts, 'tis done." That eloquent, serious-minded motto, conjured by a bunch of teenagers, shows not only breathtaking prescience, considering Armstrong's future deeds in life, but also the respect this young man already commanded. The gang at Blume High in Wapakoneta didn't

know they were walking the halls with a future ace aviator—in peacetime and in war—and the word *astronaut* wasn't yet in the general lexicon, but if they had been forced to place a bet as to who among them might become such a thing, they would have wagered on Armstrong, who, as the yearbook entry also tells us, was already building the sturdy résumé of a leader: student council vice president, home room president, Boys' State. (More surprisingly, considering Armstrong's later reputation for shyness and taciturnity: He was also in the glee club and junior year play.)

Sixteen years old in the spring of 1947, and already with the demeanor, seen by all, to think, act and get things done.

Where had this kid come from?

NEIL ALDEN ARMSTRONG was born on August 5, 1930, in the dusty outskirts of the small farming town of Wapakoneta, on Ohio's limestone prairie—115 miles north of Cincinnati. Today Wapakoneta has more than 10,000 citizens and is, with signs and murals and street names and its Armstrong Air & Space Museum, a living tribute to its famous native son. Back then, it was a place where a kid could play by the train tracks or down by the Auglaize River.

Neil's episodes of carefree fun in Wapakoneta were intermittent; his father, Stephen, was employed as a state auditor, necessitating a family move every so often to a new town. Eventually, the family, which included Neil's mother, Viola, and two younger siblings, would settle back in Wapakoneta, where Neil would excel at Blume High.

But before then, his personality would be formed, and he would develop an obsession that would prove to have not been merely a boyhood flight of fancy, but rather his life's driving force.

A smart, serious-minded child, he enjoyed reading, and so, of course, it has become part of the legend that one of the first books he ever finished was about local heroes the Wright brothers, who had, just a

ALL AMERICAN KID
In the summer of 1933 Neil turns three years old. Whether he has
yet marveled at the moon is unrecorded.

THE GOOD SCOUT
Neil at Purdue University as an Eagle Scout, 1947.

Eagle Scout

few decades earlier, started playing around with airplane prototypes at their bicycle shop down in Dayton. That story, never told by Armstrong, is hooey, as pointed out in James R. Hansen's biography, *First Man: The Life of Neil A. Armstrong.* Neil loved planes as a boy, but it was an adventure with his dad that planted the seed. One weekend when Neil was six, his father put his finger to his lips and the two of them sneaked off behind Viola's back—skipping out on Sunday school—and hightailed it to where a Ford Trimotor, known as the *Tin Goose,* was scheduled for a barnstorming stop in the area. That was Neil's first airplane ride, and he was hooked. As he grew older, he took to building his own balsa wood airplane models, and even constructed a scaled-down wind tunnel to test them.

He started learning to fly at age 14, hitchhiking to the local airport to take nine-dollar-per-hour lessons paid for with money he earned doing odd jobs around Wapakoneta, most regularly after school at the local pharmacy. On his 16th birthday, he got his license. Not his driver's license—that could wait—but his student pilot's license. He had his wings.

Since he was so dedicated to flight at such an early age, and since he went on to walk on the moon, Armstrong's story is often told in one dimension. But he was a well-rounded, well-behaved youth and adolescent. He had friends. He did well in school. He worked his way through the many requirements and earned honors as an Eagle Scout—which, any Boy Scout can tell you, is no mean accomplishment.

The Armstrongs were not poor, but Stephen and Viola did not necessarily have the wherewithal to put Neil, his sister, June, or his brother, Dean, through college. Neil, for his part, won a Navy scholarship and enrolled at Purdue University in Indiana, where he studied aeronautical engineering. When he was 19 his studies were back-burnered by the Korean War after the Navy called him up to active duty. He was sent to Pensacola, Florida, for training.

There, he lined up to fly the single-engine fighter planes. Back home, this news was nervously received, and Neil's explanation to his mother, "Mom, I didn't want to be responsible for anyone else," did little to allay her fears.

Prospective copilots he might have been responsible for were perhaps lucky, for while some aces' careers are marked by the number of kills, Armstrong's was remarkable for his number of saves. No doubt he served with distinction: 78 combat missions launched from an aircraft carrier from 1950 to 1952, three Air Medals awarded, his unit later immortalized in the novel *The Bridges at Toko-Ri* by James A. Michener. But more than once Armstrong came limping home in a crippled plane after engaging the enemy. On one occasion, he didn't make it all the way back to the ship. Having collided with an anti-aircraft cable purposely deployed by the enemy as a booby trap, his Panther Jet had nearly six feet of its right wing cut off. Armstrong stayed steady as he pointed the doomed plane back toward safety. When he was out of harm's way, he ejected, and was rescued. Years later he would add to his reputation not only for cool under fire, but in the narrower category of ejecting from aircraft at just the right moment.

After his service in Korea, he returned to Purdue and applied himself even more fully to his studies, eyeing a career in aeronautics. He graduated in 1955. Meantime, he had fallen in love with a classmate from a Chicago suburb, named Janet Shearon, whom he had known for three years before asking her out ("He is not one to rush into anything," she later told LIFE). They married in January 1956.

Armstrong had landed a job, and he and Jan would be moving west—to Edwards Air Force Base in the California high desert. Edwards was then and always will be known as the American mecca for hotshot pilots who possessed the right stuff.

Armstrong had the right stuff. His classmates at Blume High had seen that.

Where would it lead him?

THE WRIGHT STUFF
Not very far from where Orville and Wilbur built their larger models, Neil,
as a teenager, shows off one of his own planes.

A HAPPY BOYHOOD

On the opposite page, Neil is the big boy in the center, age six, pictured with his brother, Dean, and his sister, June. The kids pitched in around the house with daily chores such as weeding the garden or helping with the laundry. Above: Two years later, the boy who would become what is known in the trade as a space cowboy takes the reins. Left: Neil is second from left, playing baritone horn in the Mississippi Moonshiners, a jazz combo he formed with pals while in high school in Wapakoneta. Two of his passions in the period and later were music and flight, and at Purdue when he dated Jan Shearon, these were shared touchstones. She loved music, too, and was well familiar with airplanes because her father, a physician in Wilmette, Illinois, a Chicago suburb, had owned one for flying back and forth between Wilmette and the family summer home in Wisconsin.

Eagle Scout

SMALL TOWN LIFE

Because of Stephen Armstrong's job with the state government, the family moved some 20 times to towns across Ohio in Neil's first 15 years. But Wapakoneta, in Auglaize County, where Neil was born on his grandparents' farm and where he would finish high school, was considered home as much as anyplace was—and certainly lays claim to Armstrong as a favorite son today. Parts of downtown (above) don't look much different now than they did back when, and his grandparents' homestead on the outskirts (right) isn't far different, either. Opposite: Neil visits with his grandmother while on leave from flight training in Pensacola, Florida, where he has chosen the course of a solo fighter pilot, causing his family not a little worry.

Eagle Scout

QUICKLY, ADULTHOOD

Armstrong was pulled from college and sent to fight in Korea, where he is with fellow flyboys, third from left, above. When he returned to Purdue, his relationship with Jan Shearon took off. From an article in LIFE: "They got acquainted by running into each other around the campus in the chilly dawn. Neil had a job delivering the campus newspaper and Janet, a home economics student with a passion for swimming, had a lot of 6 a.m. lab courses. Each admired the other's industry." At right, he and Jan wed at the Congressional Church in Wilmette, on January 28, 1956. Janet subsequently would not finish her degree, something she would later regret. When the young newlyweds moved to California so that Neil could begin his career, Neil lived for a time in the bachelor quarters at Edwards Air Force Base, while Jan lived in the Westwood section of Los Angeles. They later lived in a primitive summer cabin 5,000 feet up in the mountains of Antelope Valley in northern Los Angeles County and began building a family. LIFE reported that their accommodations were "so lacking in modern conveniences that Jan bathed their first son Ricky [as Eric was called] in a plastic bathtub in the backyard after the sun had heated the water."

The Space Race

TWICE IN OUR PAGES we will pause to add context to the Neil Armstrong story: later in explaining the Apollo project, and here to set up the space race as a whole, which led to President John F. Kennedy's bold declaration in 1961 that the United States would go to the moon before the decade's end. The race was, as we all know, between the United States and the Soviet Union. It was not so much about exploration as about weaponry and military strategy. And it unfolded in a post–World War II era when bright young people like Neil Armstrong knew exactly what was going on, what might be at stake and what they might do if they were allowed to play a role.

During World War II, the Allied and Axis combatants engaged in a mortal competition to invent evermore deadly means of attack. In Europe, Germany sought to bring the island nation of Britain to her knees with assaults of unprecedented horror: destruction that arrived from the heavens with no warning save, perhaps, the eerie buzz of a flying-bomb blitz about to strike a random street corner. As the conflict wore on, the battle in the Pacific became one of methodical, costly Allied victories, culminating in the stunning denouement wrought by atomic power.

With the war in its final year, the Nazis, trapped within tightening concentric rings, desperately pursued new methods to strike from afar. The scientists in Germany had for decades shown an aptitude for rocketry—but, fortunately for democracy, the Nazis did not have enough time to produce missiles of sufficient range, potency and accuracy to seize victory from defeat. Still, by January 1945, they had launched a prototype of the first intercontinental ballistic missile (ICBM), one capable of reaching North America by traveling at an altitude of 50 miles and a speed of 2,700 mph. But again, this was only a prototype, and the war's conclusion was soon at hand. By the end of the year more than a hundred German rocket scientists were in White Sands, New Mexico, having delivered themselves to American troops, if only to avoid being taken by the Soviets.

It would be difficult to overstate the importance of these émigrés (who arrived here when Armstrong was working toward his pilot's permit in Ohio) to the U.S. space program. American rocketry, under Robert H. Goddard, had been exemplary, as German star Wernher von Braun was the first to admit: "Until 1936, Goddard was ahead of us all." But post–WWII, the U.S. was thrust into an age of guided missiles, with which it had virtually no hands-on training. Von Braun, who would become chief of the U.S. Army ballistic-weapons program, said of the German rocket men, "If we are good, it's because we've had 15 more years of experience."

The U.S.S.R. also made use of German scientists (some of them kidnapped), and while they played a smaller role, the Soviets feasted on whatever knowledge was available. For them, technology and industry were critical fields of competition in their protracted war with capitalism; furthermore, they were fully aware that American aircraft in Europe could reach their homeland, while the U.S. was as yet out of reach. Thus was sounded the communist clarion call for long-range weaponry.

American experiments during and just after the war were much more concerned with the effects of sonic speed on fixed-wing aircraft, piloted by such storied men as Chuck Yeager and Scott Crossfield—two whom young Armstrong admired from afar and later would get to know well. Their training was rigorous, their flights perilous in the extreme, but the effort produced a set of enviable successes and would coalesce with rocketry into a program of destiny—but only after that program caught up to the Soviets, who unquestionably were first off the launch pad.

Leading the early chase from Florida's Cape Canaveral were the magnificent seven of Project Mercury. Once they were successfully aloft, with Alan Shepard's inaugural mission on May 5, 1961, the competition was well and truly engaged. And then President Kennedy, only three weeks later, declared a finish line—the moon—and suddenly Neil Armstrong and a new generation of flyboys had time to enter the race.

READY, SET, GO!
When the U.S.S.R.'s Sputnik 1 streaked across the sky (here, over Montreal) in October 1957, it was a shot across the bow to the U.S. The first artificial earth satellite was a surprising success, and in its elliptical low planetary orbit, instantly threatening. The Space Age was upon us. The Space Race was soon to follow.

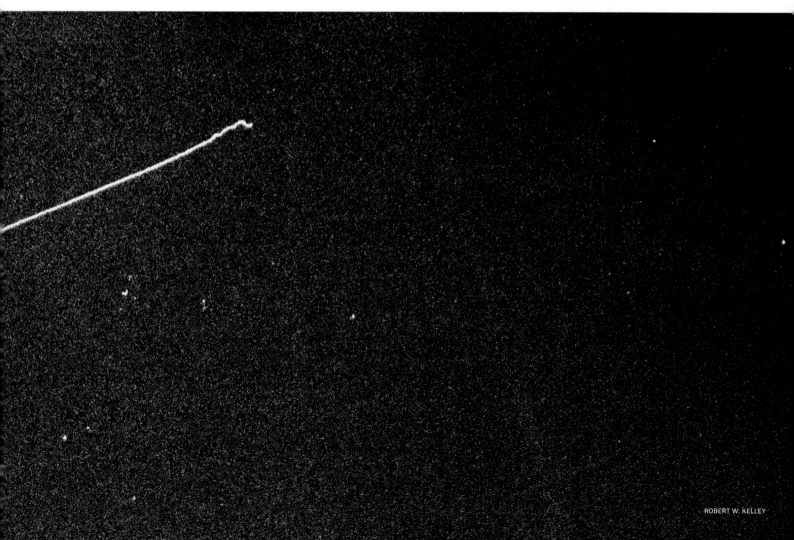

The Space Race

LIVIN' TO FLY, FLYIN' TO LIVE

In the 1950s, before the space program got going in earnest, the hottest of the hotshot flyboys found their way to Muroc Army Air Base in California's Mojave Desert—it would later become Edwards Air Force Base—and raise hell, both on and well above the ground. Above is Chuck Yeager, still today as close to a living legend as America has. On October 14, 1947, he and his two broken ribs climbed into a dangerous, rocket-powered beast—an X-1 named the *Glamorous Glennis,* after his wife—and he became the first man to break the sound barrier. Opposite: Meet the men of Project Mercury. These photos were all taken by Ralph Morse for LIFE, which had registered a great—if controversial—journalistic coup when it agreed to pay the astronauts some $70,000 each to showcase their roles. When the first of a series of lengthy articles kicked off (that's the cover), the glimpse into the training and homes of America's newest heroes was galvanic, inspiring NASA chief James Webb to thank LIFE for "waking up the country to the space program." Rival publications chafed at the exclusivity of the deal, and NASA itself derided the commercialization of the program—but they had given it the green light. Finally, even the managing editor of *The Washington Post* conceded that the stories' success relied on hard work more than favored treatment.

23

The Space Race

THE MOON IN THE CROSSHAIRS

Read today in an equivocal age, President John F. Kennedy's "Man on the Moon" address, delivered to Congress on May 25, 1961 (below), is a remarkable artifact. The raison d'être for the space race, as we vaguely remember it—that it was about the cold war at least as much as it was about science, adventure, exploration or even conquest— is baldly stated by the President, with no ifs, ands or buts about it. In soliciting billions of dollars for the effort, in "asking the Congress and the country to accept a firm commitment to a new course of action—a course which will last for many years and carry very heavy costs," Kennedy stressed that the outcome of the space race would impact "the minds of men everywhere, who are attempting to make a determination of which road they should take." These roads, said Kennedy, headed in the diametrically opposed directions of "freedom and tyranny." The speech is rife with concerns that the United States is lagging badly, and that space "may hold the key to our future on earth." The Chief Executive declared: "I believe that this nation should commit itself to achieving the goal, before this decade is out, of landing a man on the moon and returning him safely to earth . . . I believe we should go to the moon." Toward that end, Alan Shepard becomes America's first man in space (right) and, in the three photos on the opposite page, John Glenn becomes the first to orbit the globe, then is congratulated by his Mercury mates.

KINGS OF THE X-15

Armstrong's smiling (above), but this time during reentry from 207,500 feet in his X-15, he inadvertently established a positive angle of attack during pull-out, and overshot Edwards Air Force Base heading south at Mach 3. He finally managed to turn back near Pasadena, California, and had just enough energy to land on the south end of Rogers Dry Lake at Edwards. Opposite: That kind of save certainly would have been appreciated by X-15 legend Scott Crossfield (left) who ceremoniously hands over the keys to the rocket plane to Armstrong as Major Robert White looks on.

"One Generation Late"

NEIL ARMSTRONG was in his freshman year at Purdue when the big bang of '47 was heard: Chuck Yeager blasting through the sound barrier. A devotee of aviation, Armstrong paid attention, and surprisingly found himself a little sad. Not that he was sufficiently presumptuous while still just a teenager to think that it could have been him piloting that rocket-powered Bell X-1, but he did grow winsome. The Wright brothers of Dayton, Charles Lindbergh, Richard Byrd, Amelia Earhart, the combat aces of World War II, and now Yeager and Scott Crossfield and those other boys out in the California desert . . .

He later recalled to his biographer James R. Hansen, "All in all, for someone who was immersed in, fascinated by and dedicated to flight, I was disappointed by the wrinkle in history that had brought me along one generation late. I had missed all the great times and adventures in flight."

It's funny: Most of the Mercury 7 astronauts felt the same way. They were real fliers, like Yeager, and some had stared down death during combat, as Armstrong had. (An irresistible example: John Glenn flew 59 combat missions as a kid in World War II and another 90 over Korea, where he earned the nickname "Magnet Ass" for his ability, shared with Armstrong, to attract enemy fire after not once but twice returning to base in a plane that was riddled with bullet holes.) These were guts-and-glory guys, faster-than-sound test pilots, and this man-on-a-candle stuff, with mission

control calling the shots, hardly seemed like real flying to them. They recoiled, and were grudgingly granted some liberties at the wheel by their NASA overlords.

Little did anyone know, back when it seemed like the aviator's golden age was drawing to a close and the age of the "astronaut" was dawning, that in the very near future a premium would once again be put not only on courage but on skill in the outer-space program, and that such men as Jim Lovell and Neil Armstrong would gain renown for flying feats performed not above or just beyond earth's atmosphere, but a world away—in the lunar sphere.

NEWLYWEDS Neil and Jan arrived in California in 1956 and Neil went to work at Edwards Air Force Base as an experimental test pilot for the National Advisory Committee for Aeronautics. He was a direct heir of Yeager and Crossfield, and also a bridge to the future. A decade before, Yeager had taken off from these exact same runways in the Mojave Desert. (The base was called Muroc at the time; it was renamed in 1950 for Glen Edwards, a test pilot who had been killed in a crash of his Northrop YB-49.) Crossfield had been the first to conquer Mach 2 (twice the speed of sound) and would be the first to pilot the X-15. Now here was Armstrong, taking his first flight in a rocket-powered plane in the Bell X-1B, a successor to the Yeager craft, and soon to take the X-15 to speeds in excess of 4,000 mph seven times, reaching

"One Generation Late"

and even exceeding the very ceiling of the earth's atmosphere. "The most technically capable of the early X-15 pilots," was fellow pilot Milt Thompson's assessment to *The New York Times*. Marveled their teammate Bill Dana, "[Armstrong] had a mind that absorbed things like a sponge and a memory that remembered them like a photograph."

So Armstrong was a top-notch test pilot right out of the mold. But the National Advisory Committee for Aeronautics was soon to morph into the National Aeronautics and Space Administration, and the focus of this new NASA was shifting east from Edwards to the training facilities in Houston and the launching pads of Cape Canaveral in Florida. The Mercury 7 were well on their way to their first missions, but as the 1950s segued to the '60s, Armstrong was all of a sudden faced with an interesting choice: He was offered the chance to test a prospective military space plane, and meantime NASA was accepting applications for a second team of astronauts. He had once discounted the authorized space program, believing the winged X-15 approach stood a better chance up there than a capsule did. But John Glenn's successful orbital journey had been persuasive, and now, as Armstrong later recalled to Hansen, "I thought the attractions of being an astronaut were actually, not so much the moon, but flying in a completely new medium."

Nineteen sixty-two was as tumultuous as (and in one way much more tragic than) any imaginable spaceflight for the Armstrongs. Neil and Janet's young daughter, Karen, died of an inoperable brain tumor. That of course was the crucible. Then, too, on April 20, Armstrong, zooming up to 207,500 feet—well above the limits of earth's atmosphere—in the X-15, saw his plane behave as it hadn't in simulations, and had to wrestle it to a harrowing landing. Also, the family, which included son Eric (known as Ricky), packed its bags for Houston (where son Mark would be

born in '63). And finally, Dad started his new job as the first civilian chosen to be an astronaut—a veteran working alongside fellows who were still in the military.

There was little hazing, as Armstrong quickly proved, as he always did, his considerable worth. He was assigned to the Gemini program, which would see a series of launches of two-seat capsules, improving on Mercury's solo missions and pointing the way to the three-astronaut Apollo flights that would take mankind, finally, to the moon. Armstrong progressed apace, was a backup on the Gemini 5 mission and then was assigned to be commander of *Gemini 8,* with David R. Scott as his copilot.

Neil Armstrong was still not on the public's radar screen, but if anyone had bothered to inspect his now considerable career they might have expected that if something were to go wrong up there, Armstrong would come through. On March 16, 1966, Armstrong and Scott had successfully engineered the first ever docking of two space vehicles, a maneuver that would be crucial if Apollo missions were to make it to the moon. So there they were, locked up with the unmanned *Agena* spacecraft, when the conjoined whole began to roll. Mission control told Armstrong to disengage *Gemini* from *Agena,* which he did, but the tumbling worsened. The problem was in the capsule (in point of fact, one of its thrusters was firing continuously). The situation was quickly getting to the point where the astronauts might pass out. Armstrong and Scott shut down the flight control system governing the thrusters, and switched to reentry control. The capsule stabilized. The flight was aborted at that point, and the crew was instructed to return home, not having finished a day in orbit.

Armstrong, back on terra firma, was undeterred as he transitioned from the Gemini to the Apollo project. He was now becoming something of a legend among his peers. If you got into trouble with this guy, somehow you'd get out of it.

WELCOME TO HOUSTON . . .
. . . and to the astronaut corps. In the summer of 1963, Neil and Jan pose with their boys Ricky (standing) and Mark after Dad's career shift from planes to capsules. In a way, this photograph signifies a rite of passage, as it was made for LIFE by Ralph Morse. He was our man on the space beat from the first, and enjoyed such access to the early American space superstars that he was jokingly known as the Eighth Mercury Astronaut, in the way of all of those later claiming to be the Fifth Beatle. Among Morse's many famous pictures of the space program were the frames in his sequence of the *Apollo 11* launch, which we will see later in these pages.

"One Generation Late"

VETERANS AND ROOKIES
In Houston in 1963, the original Project Mercury astronauts pose with new trainees for the Gemini and Apollo programs. Seated, from left: Gordon Cooper, Gus Grissom, Scott Carpenter, Wally Schirra, John Glenn, Alan Shepard and Deke Slayton. Standing, from left: Ed White, James McDivitt, John Young, Elliot See, Charles Conrad, Frank Borman, Neil Armstrong, Tom Stafford and Jim Lovell.

BETTMANN/CORBIS

KEYSTONE/GETTY

IT'S GOING TO BE LIKE THIS?

In training sessions in 1963, Armstrong and his colleagues prepare for conditions that, quite frankly, they are unlikely to confront in space. But certainly, they are being toughened generally, and that will serve them well. Left: Armstrong peers from a makeshift tent in Nevada during an intensive three-day desert training course. Above, in Panama, the astronauts' fare is a "jungle" meal as they take part in tropic-survival training at Albrook Air Force Base. Wally Schirra is taking a piece of wild pork from a stick held by Jim Lovell as, from the left in front, John Glenn, Scott Carpenter and Deke Slayton chow down. Behind Schirra are Gus Grissom, Neil Armstrong, Tom Stafford and Alan Shepard. The truth is, Armstrong was never much for heavy training. He told LIFE that he figured a human being was given a life's quota of heartbeats, and he didn't want to "waste" any heartbeats on exercise.

"One Generation Late"

IT COULD BE LIKE THIS
Front, from left: James McDivitt, Ed White and Neil Armstrong participate in a weightlessness drill. A fun anecdote at this point: When Armstrong was a small boy, he had a recurring dream that he could, by holding his breath, hover over the ground. Nothing much ever happened—he neither flew nor fell, just hovered. As delightful as the dream may seem to us, it could be frustrating to a child as captivated by the idea of flight as young Neil was. "I tried to do it later, when I was awake," he once told LIFE with a shy smile. "It didn't work." And now, in 1963, thanks to NASA: It does.

NASA/PHOTO RESEARCHERS/GETTY

"One Generation Late"

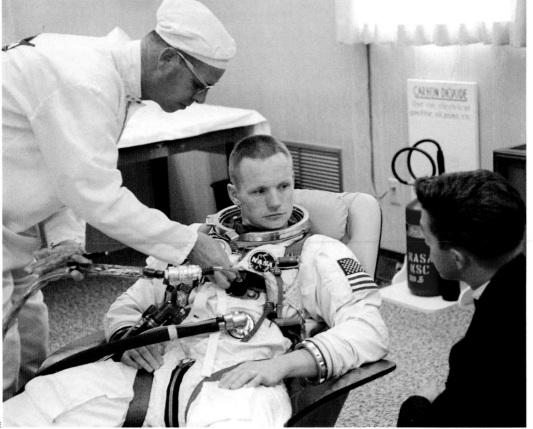

BETTMANN/CORBIS (2)

AP

ASSIGNMENT: GEMINI 8
Armstrong's first shot at space would come in the 1966 Gemini 8 mission alongside David Scott, who would also go on to walk on the moon as a member of the *Apollo 15* crew. Scott is at left and Armstrong at right in two of these photos (the situation above is during a simulated test), and Armstrong is being checked out in the third photograph (left). Both astronauts would be rookies on this flight, and what a wild ride they would take. Just incidentally, Armstrong, who had retired from the Navy in 1960, would become the first civilian NASA astronaut in space on *Gemini 8.*

"One Generation Late"

CALM BEFORE THE STORM

All is peaceful as Command Pilot Armstrong maneuvers the *Gemini 8* capsule toward the *Agena* rocket, seen here, in preparation for docking on March 16, 1966. It is only after the two craft engage that things go wrong, and the joined vehicles start to tumble in space. Later, both pilots will be credited for their performances in saving the bacon, and it will take time for the true severity of what had happened to be learned. Neil's wife, Jan, could not have been aware at the time just how critical the situation was, but as she watched televised accounts of the mission she was overwhelmed when the flight was aborted and an emergency landing in the Pacific ordered. These pictures were the product of a tradition at the time: that LIFE would be invited into the family home to take pictures of the wives as their husbands traveled through space. Sometimes, certainly, when things grew tense, our presence was intrusive.

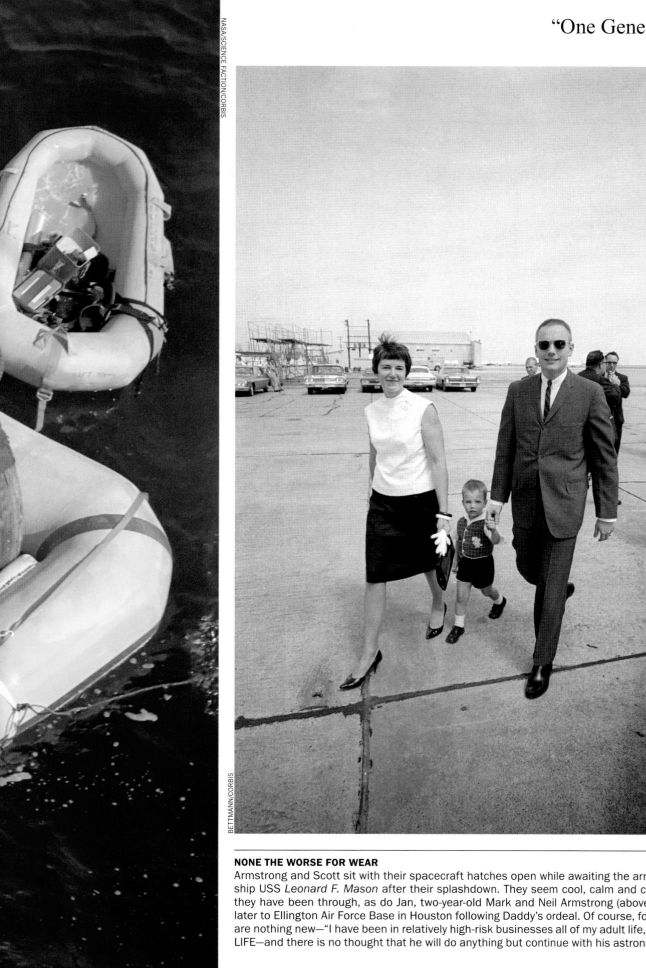

NASA/SCIENCE FACTION/CORBIS

BETTMANN/CORBIS

NONE THE WORSE FOR WEAR
Armstrong and Scott sit with their spacecraft hatches open while awaiting the arrival of the recovery ship USS *Leonard F. Mason* after their splashdown. They seem cool, calm and collected despite all they have been through, as do Jan, two-year-old Mark and Neil Armstrong (above) when they return later to Ellington Air Force Base in Houston following Daddy's ordeal. Of course, for Armstrong, crises are nothing new—"I have been in relatively high-risk businesses all of my adult life," he understated to LIFE—and there is no thought that he will do anything but continue with his astronaut career.

Apollo

ANOTHER BRIEF INTERLUDE HERE, as we forward the Neil Armstrong story.

The Apollo program of which he would become the most prominent face was begun when Armstrong was still a flyboy in the Mojave Desert thinking the glory days of aviation were in the rear-view mirror. Apollo did not, in fact, start when Project Gemini wrapped up in 1966. It started back in 1961 when John F. Kennedy said we were shooting for the moon. Gemini was a concurrently running program, always in service to Apollo, and actually Apollo had been considered back in the 1950s, during the Eisenhower Administration, as a sequel to Project Mercury. Kennedy put the whole big train on the rails with his moon promise, and it had rolled on throughout the first half of the decade—Saturn rockets and lunar modules and three-person command modules being developed and refined even as Mercury figured out launches and reentries and Gemini worked on out-there stuff such as docking and space-walking.

Apollo was ready for its manned debut in 1967 when it—and the entire space program—suffered a disaster unlike any it had ever experienced.

As Armstrong's thrilling career implies and as we all know intuitively: The business of pushing aircraft to their limits is just about as precarious a human endeavor as can be imagined. There were flight-related deaths nearly every year at Edwards Air Force Base when the astronauts we came to know were testing planes there—in one year, 14 fatalities—and if the rockets and capsules of the space programs had so far escaped unscathed beyond a few harrowing reentries and Armstrong's tumble aboard *Gemini 8,* well, that was about to change. Kennedy himself, obviously a booster, had warned in 1963 that space travel represented "the most hazardous and dangerous and greatest adventure on which man has ever embarked." His words, four years later, would be borne out.

The launch of the prospectively named Apollo 204 mission was targeted for February 21, 1967, and all seemed to be going well, although later investigation proved there were any number of design flaws in the command module, and that there had been concerns voiced by the astronauts themselves, in particular Command Pilot Gus Grissom. In an anecdote that is now part of Space Race lore, the legendary veteran of Mercury and Gemini missions was on his last visit home on January 22 when he pulled a lemon off a tree in his yard. His wife, Betty, asked him what he was going to do with the fruit. "I'm going to hang it on that spacecraft," he answered. He kissed Betty goodbye and headed east.

Three days later, a short circuit during a mock launch sequence aboard what would later be renamed *Apollo 1* set off a flash fire that killed the three-member crew, who are seen here. Only three months later, Soviet cosmonaut Vladimir Komarov died when his capsule's parachute lines tangled in reentry. He, Yuri Gagarin, who had been the very first man in space in 1961, and others had protested that the new *Soyuz* spacecraft wasn't yet flightworthy, but Premier Leonid Brezhnev ordered the launch. Komarov was still cursing as he crashed. A rumor maintains that when Gagarin next saw Brezhnev, he threw a drink in his face.

Suddenly, everyone was questioning the Space Race—its validity, its cost, whether it was moving too fast. It was 1967, the Vietnam War was raging, the counterculture was at full cry, and the glamour and heroism of such as John Glenn seemed yesterday's news.

NASA was on its heels. It launched its *Apollo 1* investigations as well as a series of unmanned Apollo missions (it would be 20 months before it would return to manned flight with *Apollo 7* and then *8,* the latter of which became the first manned spacecraft to orbit another celestial body). Slowly, interest in Kennedy's pledge returned, and then it became manifestly clear that the space program was doing all it could to make good on JFK's promise by decade's end. The buzz was back.

We will revisit the global excitement caused by *Apollo 11* in our next chapter.

In all, by 1972, the Apollo program would send seven missions aloft with the intention of landing on the moon. Six would succeed and there would be

one more great, harrowing drama when Commander Jim Lovell's *Apollo 13* suffered an oxygen tank explosion that disabled the command module's propulsion and life support, forcing the crew to use the lunar module as a so-called lifeboat to return from the dark side of the moon. Twelve of Lovell's colleague astronauts—including America's first man in space, Alan Shepard—would eventually walk on the lunar surface.

None have done so in 40 years.

ILL FATED
A flash fire during a test on *Apollo 1* in 1967 claimed the lives of, from left, Gus Grissom, Ed White and Roger Chaffee. Theirs was the space program's first great tragedy, but, with two space shuttles having been lost since, not the last.

Apollo

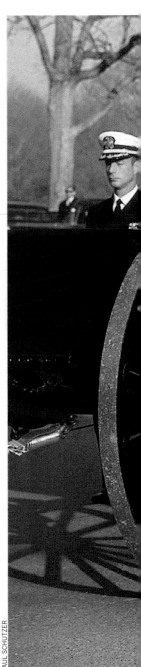

FIRST TASTES OF TRAGEDY

On the page opposite, Gus Grissom is seen holding a notepad in the foreground of the top photo, along with other Apollo astronauts, and at right in the bottom photo alongside his *Apollo 1* crewmates, Roger Chaffee and Ed White. Chaffee and White were well-known, certainly—White had been the first American to walk in space—but Grissom was a legend: decorated Korean War fighter pilot, second American in space, during Project Mercury, veteran of Gemini as well, about to launch the Apollo program. Because of him, and the subsequent report that he had had qualms about the capsule's safety, the tragedy was enormous, scandalous news. Above: Fellow Mercury veterans (from far left) Scott Carpenter, John Glenn and Gordon Cooper attend Grissom's casket as taps drifts over Arlington National Cemetery, outside Washington, D.C.

Apollo

RETURN TO MANNED MISSIONS

It fell to the crew of *Apollo 7* to try to restore luster to the space program after a 20-month hiatus between manned flights. They were, left to right in the picture at left, Donn Eisele, Wally Schirra and Walter Cunningham, with Director of Flight Crew Operations Deke Slayton standing. They certainly tried their best to revive the spirit, even unto posing for the goofy photographs (top, from left, Cunningham, Eisele and Schirra), but their flight is perhaps best remembered for the cold Schirra caught while aloft, which he passed to Eisele. The astronauts grew grumpy with ground control during their orbits, and maybe it is not coincidental that none flew for NASA again (though it should be noted, Schirra was the first to fly in the Mercury, Gemini and Apollo programs). Houston was nice enough to send U.S. Navy frogmen to fetch the astronauts after their splashdown near Bermuda (above).

Apollo

A TRIP TO THE MOON

If it was difficult for the Apollo 7 program to recapture the public's imagination, it was much less so for Apollo 8, because this entailed something entirely new: a trip around the moon. Having launched on December 21, 1968 (opposite), Commander Frank Borman, Command Module Pilot Jim Lovell and Lunar Module Pilot William Anders took three days to travel to the moon, which they orbited 10 times in 20 hours. Their Christmas Eve televised reading from way up there of verses from Genesis was, at the time, the most-watched TV event ever. Then it was time to head back home, having proved that, perhaps, this man-on-the-moon thing could be done after all. On December 27, when the spacecraft splashed down in the northern Pacific Ocean, there was relief in Washington, where President Lyndon Baines Johnson sipped tea and followed the news (above), and elation at mission control in Houston (right), where Neil Armstrong (left) joined others in celebration, if not in cigars.

Apollo

MOONSCAPE?
We've seen NASA training trips to the desert and the jungle, and now we have Iceland, which was visited by American astronauts first in 1965 and again, here, in 1967. The popular theory is that this was because the Icelandic terrain resembled what was expected of the moon's topography, and if there were ancillary benefits, good—but these were primarily geologic field trips. Nine of the 12 men who would eventually set foot on the moon between 1969 and 1972 visited Iceland on one or other of the two trips. Neil Armstrong did, arriving on July 5, 1967. He enjoyed his sojourn, sneaking off at one point with fellow astronaut Bill Anders for fishing on the River Laxá.

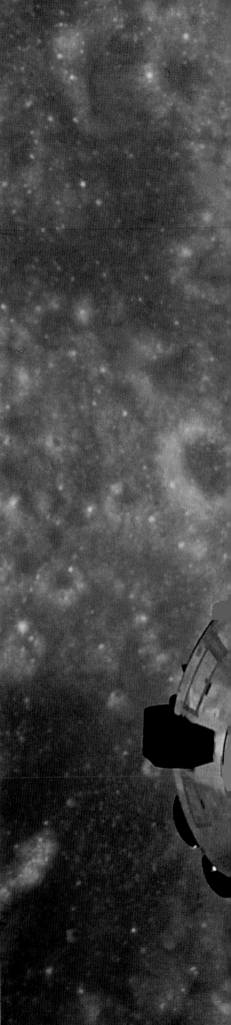

SSPL/GETTY

CLOSER AND CLOSER

Launching on May 18, 1969, *Apollo 10* was the fourth manned craft in the program and the second to orbit the moon, this time with an all-up test of the lunar module while in orbit. The module came to within 8.4 nautical miles of the lunar surface during practice maneuvers. Above we see astronaut John Young shaving while Tom Stafford looks on; Gene Cernan is also in the command module, which has been nicknamed Charlie Brown. That's Charlie from the outside (right) in a photograph taken from the lunar module, which naturally has been nicknamed Snoopy. And why? Because the creator of the *Peanuts* comic strip, Charles Schulz, is a big NASA fan, and has drawn special artwork to promote the mission back on earth.

NASA

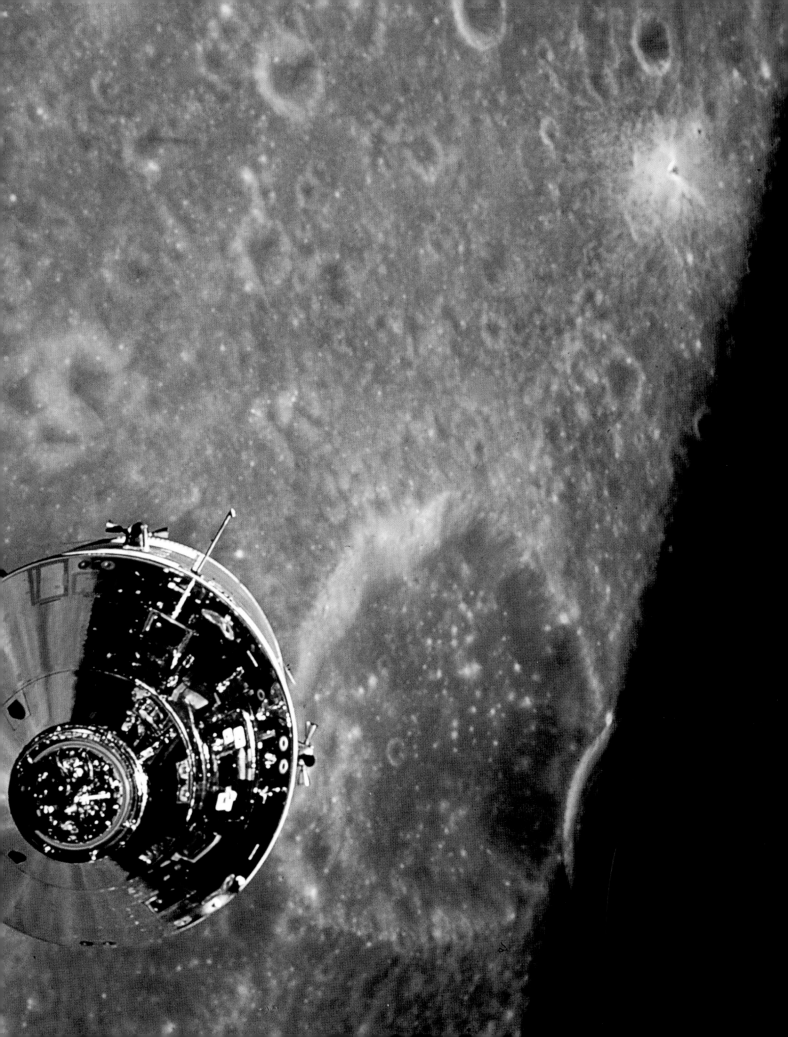

To the Moon and Back

YOU TRAIN YOUR WHOLE LIFE for a thing, and unless you're working solo, there's no guarantee—the opinions of others, the talents of teammates and the vicissitudes of fate are always at play. Edmund Hillary, who in 1953 was certainly one of the world's best mountain climbers, was a New Zealand beekeeper, and there is no way he would have reached the summit of Mount Everest on his own. But there was an elaborate and well-financed British expedition trying to conquer the world's highest peak that year (think of it as the space mission of its day); New Zealand was a Commonwealth country and so Hillary technically qualified as a candidate; Hillary had the good sense to choose the swift and strong Sherpa Tenzing Norgay as his climbing partner; John Hunt, the expedition leader, had the good sense to recognize Hillary and Tenzing's abilities in the field; and so those two were tabbed for the final assault. Did sponsors wish, back in London, that Hunt had not picked a Kiwi but rather a dyed-in-the-boiled-wool Englishman from his deep team of climbers? No matter. Hillary had the right stuff and so did Tenzing, and so they are the men in the history books.

A crucial extra point to be made, though: It wasn't all luck.

They had put themselves in position, and they were ready when called.

And so were Neil Armstrong, Buzz Aldrin and Michael Collins, who certainly had to succeed at several rolls of the dice before they became the chosen ones to reach the moon—which represented in 1969 the most notable attempted "conquest" since Hillary's 16 years earlier.

To return to this question of chance for a moment: What if *Apollo 1* hadn't suffered its terrible fate? The shuffling of the astronaut deck as mission after mission launched surely would have been different. (Not least, several more of the early Apollo missions would have been manned, and maybe the Armstrong crew would have been used up.) What if there had been

additional setbacks or, God forbid, tragedies? Might the moon shot have been abandoned altogether if things hadn't gone well for *Apollos 7, 8, 9* and *10*?

What if, for that matter, Armstrong had been killed in the Korean War? Or when his test plane ballooned out of control above the hardpack of high-country California? Or when he and David Scott went ass-over-teakettle in *Gemini 8,* the capsule spinning at a full revolution per second, Armstrong confessing to ground control, "Well, we've got serious problems here," then confessing to his dizzy crewmate, "I gotta cage my eyeballs."

Or when he lost that lunar landing training vehicle.

Ah, yes: another Armstrong episode, another ejection from another flying machine, this time with a $2.5 million NASA toy lost in the bargain, but Armstrong himself still in one piece—but of course—having sought safety, in this instance, with perhaps 1.5 seconds to spare. What had happened: In 1968 Armstrong, now with his focus set squarely on the moon per NASA's announced two-year manifest, was piloting a trainer based on the lunar landing module, and suddenly black smoke was pouring forth—not a good sign when he was driving a paste-job jalopy in Ohio back in the 1940s, not a good sign now. Two hundred feet above the ground, the thing started spinning, and goodness knows Armstrong was still sick of spinning. He tried to bring the craft in, but realized this was a lost cause, ejected, and floated to the ground with his parachute deployed. After the *Gemini* incident with Scott, an observer on the ground had reported two almost shockingly calm astronauts, considering the ordeal. And now a NASA spokesman, when asked by reporters what the heck had gone on during that training session, said Armstrong was fine, he was "back in the hangar, walking around, discussing the incident." This was the episode alluded to in our book's introduction, where Armstrong is asked if

IN SYNC?
Buzz Aldrin (left) and Neil Armstrong are in training for what is to come in the lunar landing module, in which they will descend to the moon's surface. They will be standing much like this, legs comfortably apart at shoulder width, monitoring transmissions from Houston and able to see outside through small triangular windows. At this point during training, there is lingering tension between the two men. Aldrin wanted and expected to be the first man on the moon, and it seems that honor will go to Armstrong. Crewmate Michael Collins would later remember Aldrin ragging on Armstrong after a simulation exercise, and thought the big decision might be the cause.

RALPH MORSE

To the Moon and Back

the 1.5-second estimate is about right, and he responds, "Yeah."

ARMSTRONG WAS TO BE the commander of *Apollo 11.* Why? He was going to be the first to set foot on the moon. Why?

Here we arrive at the subject of the many myths surrounding this historic mission. Let's get rid of the easy ones first. The as-if-scripted drama of the lunar landing and then the weirdness of the film footage from the moon's surface surely means that the whole thing was put together on a Hollywood soundstage, with exterior scenes shot in the Nevada desert. Well, that would be false. Why would the world be caring so much about the passing of Neil Armstrong 43 years later, and for that matter: With five subsequent lunar landings, not one astronaut has come forth to rat out NASA?

Here's a fun myth: Armstrong muttered, "Good luck, Mr. Gorsky," during his moon walk because, many years earlier in Ohio, a neighbor's wife had, in the heat of an argument, screamed at her husband that he would next get lucky "when the kid next door walks on the moon." It seems that this one was a crowd-pleaser made up by the popular comic of the time Buddy Hackett.

On more serious notes, we will deal with what Armstrong did or didn't say when stepping foot—"a" or not "a"—in just a moment, and right now we deal with: Why him? A resilient suspicion is that the White House insisted, for symbolic purposes, that the first man on the moon be a civilian rather than a military man. An immediate response is: Why would they insist on that, even in the Vietnam era, since part of the goal remained to face down the Soviets? A second response is, don't be silly. Much too much was at stake to entrust the commandership of this mission to anyone but the most qualified.

But (goes the follow-up question): Would the commander necessarily be the first down the ladder?

Well, that's a point. During Gemini, it was the copilot who got to space walk, arguably the second coolest thing in the galaxy after being the first earthling on the moon. Why was it decided Commander Armstrong would do the honors, especially since early reports had it that his teammate in the landing craft, Buzz Aldrin, was slated to be the one?

This will never be sorted out for sure, but it seems to have come down to that elusive notion "character"—not Aldrin's certainly, but Armstrong's in particular. All those close calls, all those deft escapes, all that insouciance in the aftermath—all those "it was nothing" postmortems . . . And this, on top of a seriousness of purpose and a recognition that this was a big thing indeed. Armstrong wasn't only the right man at the right time. He was the right man.

Practical reasons were given for what might have been a foregone decision: The inward-opening door of the lunar module meant that the commander was in a more

convenient position to exit first. Of course, Aldrin would have been happy to scoot over Armstrong and has admitted as much. As Kathy Sawyer wrote in a fine look at Neil Armstrong's career published in *The Washington Post Magazine* in 1999: "His professionalism made Armstrong a natural choice for Apollo 11 mission commander, but a preliminary checklist had Aldrin stepping first out the lander door, and he lobbied hard to keep that position . . . Astronaut Pete Conrad, who . . . trained with Armstrong in the Gemini and Apollo programs, said the issue 'never came up until Buzz brought it up.' Armstrong just let Aldrin's complaints wash over him and allowed Deke Slayton, who ran the astronaut office, to handle the matter, according to Conrad."

Sawyer points her readers to Aldrin's memoir, in which he wrote that he did bring the matter up with Armstrong, and that "Neil hemmed and hawed for a moment and then looked away, breaking eye contact with a coolness I'd never seen in him before. 'Buzz,' he said, 'I realize the

CREW AND SUPPORT CREW

Opposite, from left: Armstrong, Collins and Aldrin are, in 1969, NASA's latest golden boys, hoping to restore luster to a space program buffeted, not so long before, by tragedy. Above, clockwise from lower left: The Aldrins, the Collinses and Jan and Neil Armstrong with their two boys. Since the beginning of the Mercury program, portraits like the ones on these pages, often appearing first in the pages of or on the cover of LIFE, served to bolster public support for the space program.

To the Moon and Back

THE CLOCK IS TICKING
Below: Collins, Armstrong and Aldrin share a drink and a moment—a light moment, thank goodness, between Armstrong and Aldrin. Bottom: Collins, Aldrin, a Mission Control engineer and Armstrong discuss details, quite rightly boning up on what's expected of them with several issues of LIFE. Opposite: Armstrong, characteristically, is focused, which will serve him and his crewmates well.

historical significance of all this, and I just don't want to rule anything out right now.'" Then Sawyer references the memoir of the third *Apollo 11* crewman, Michael Collins, who remembered an incident during training. "Aldrin was drinking scotch," writes Sawyer, "and complaining loudly about Armstrong's having crashed and burned, figuratively, earlier that day during a simulation—a kind of dress rehearsal for the lunar landing. Armstrong, 'in his pajamas, tousle-haired and coldly indignant,' confronted Aldrin. Collins speculates that what really triggered the fight was Aldrin's pique over Armstrong's exercising 'his commander's prerogative to crawl out first' on the moon. But Armstrong, years later, during one of the obligatory Apollo anniversary press briefings for NASA, told reporters flatly that, whatever his crew mates might think, he had 'zero input, no input whatever, into that decision.'"

Looking at the man and his now completed 82-year life, that sounds about right.

ONCE THE DECISIONS and dramas on the ground had been sorted out, it was time for liftoff. Under Armstrong's leadership, the *Apollo 11* trio was, in the days before the launch, all business—confident, calm. "The quietest crew in manned space flight history," said an associate. In Houston, the three family men all enjoyed quality time with their wives and kids. If the men shook off the tension, not everyone could. "I want him to do what he wants," Joan Aldrin, Buzz's wife, told LIFE, "but I don't want him to."

It seemed suitable that, long before the liftoff of *Apollo 11,* Cape Canaveral had been renamed Cape Kennedy in honor of the late President, who had reenergized the space program and set the goal. (The Cape reverted to its Spanish name, which means "place of reeds," in 1973.) Before dawn illuminated July 16, 1969, half a million spectators and more than 3,000 journalists filled the nooks and crannies around the Space Center to watch the historic event.

Armstrong led his team from the gantry to the massive Saturn V rocket that would boost his command module, *Columbia,* into space. At 9:32 a.m., the ground shook when 7.5 million pounds of thrust powered the 3,050-ton *Apollo-Saturn* stack off Pad 39A. In short order, the spacecraft broke free of earth's gravity, then spent three days making its way through space to lunar orbit. There, its fragile module bearing two astronauts would be released and would aim for yet another aptly named site: the Sea of Tranquility.

Three days after liftoff, Commander Neil Alden Armstrong of the U.S. Navy and Purdue University (not to mention Blume High); Edwin Eugene Aldrin Jr., a former Air Force pilot with a Ph.D. from M.I.T. in aeronautics and astronautics; and Michael Collins, a former Air Force pilot—each of them born in 1930 (and the other two surviving Armstrong in 2012)—were hundreds of thousands of miles from home, circling the moon. The amazing drama was unfolding.

Armstrong and Aldrin climbed aboard

To the Moon and Back

the lunar module they had named *Eagle*. Their compartment was nine feet high, 13 feet wide and 14 feet long and, besides its communication center, contained nothing more than the men, a guidance computer and some food and water. This wasn't really about scientific inquiry, this was about getting there: *Eagle* had been designed for lunar landing, lunar liftoff and ultimate rendezvous with the command and service module (CSM), manned in orbit by Collins during his teammates' absence.

On Sunday, July 20, the lunar module broke from the CSM. "The *Eagle* has wings," Armstrong said, as he and Aldrin started to make their way the 69 miles to the landing site. The phrase, so elegant and poetic and so unexpected from the taciturn Armstrong, probably raised eyebrows back in Houston, even as it indicated that the commander had put some considerable thought into the importance of what was occurring. Before the mission had begun, flight director Clifford Charlesworth had jokingly imagined Armstrong's big moment kicking the lunar dust: "I imagine he'll call Houston and say: 'We've landed.'" Now, surely, he had no idea what his man in the sky might come up with.

To accompany the new gravitas of Armstrong's elocution, there was quite suddenly gravitas to Armstrong's situation— yet another real Neil Armstrong episode. He and Aldrin were standing at their controls just more than six miles above the lunar surface when alarms started sounding. Houston reported that their onboard computer was overloading, which caused Aldrin to feel "that first hot edge of panic." With a mile left to go in the descent, Armstrong assumed partial manual control. Suddenly, he saw below him a "football-field-sized crater with a large number of big boulders and rocks." He anxiously started working the stick and the toggle to avoid where the computer was sending him. He kept the module on the move even as it kicked up dust from the lunar surface: "It was a little bit like landing an airplane

when there's a real thin layer of ground fog, and you can see things through the fog. However, all this fog was moving at a great rate, which was a little bit confusing." And he was spending fuel. Aldrin, monitoring the radar, reported their progress: four hundred feet from touchdown . . . forty feet. Houston told the astronauts they were down to 50 seconds of fuel. Armstrong brought the module down; he had found a good place. His heartbeat was 150 a minute as he reported, "Houston, uh . . ." He gathered himself, "Tranquility Base here. The *Eagle* has landed."

Touchdown was the big moment for Armstrong—"a real high in terms of elation. It marked the achievement that a third of a million people had been working for a decade to accomplish."

Many millions more were intently monitoring it as well. Several hours after the two astronauts had landed on the moon and made their preparations, a world watched rapt as shadowy images, beamed back through space, showed a man descending a short ladder and stepping onto the moon. What did he say? Did you hear what he said? What Armstrong said—"That's one small step for a man, one giant leap for mankind"—seemed perfect at the time, even if some scientists, philosophers and the man in the street would debate in the years ahead how, precisely, getting to the moon represented a leap for mankind, not to mention whether Armstrong had actually said "a."

That seemed to be the bigger controversy: Did he say "a man" or just "man": One small step for *a* man, or one small step for *man*. It seems certain, with everyone asking him about it after the fact, that he had meant to say "a man"—and indeed might have said it, only to have it garbled in a nanosecond transmission glitch. He surely had thought about what sentiment he wanted to express in the six and a half hours in the lunar module between touchdown and descending the ladder; he had earlier told his mother that he wanted to

TREASURED TIMES
Above: Neil pitches to Ricky and, opposite from top left, cooks up pizza for the family, offers (or forces) a smile and teams with Jan to explain to the boys what is in the offing. Certainly the parents do not share everything about just how dangerous Dad's latest adventure might be. Neil surely realizes the dangers, and realizes as well that, but for good fortune and his own instincts for survival, he might already have been taken from the picture. He realizes he has gone through, and could go through again, what is depicted in the photographs on the very next pages.

come up with something that included everyone in the whole world, and what is so beautiful about the line is that, at the culmination of a mammoth battle in the cold war, there is nothing the least bit jingoistic, nationalistic or even patriotic in it. Clifford Charlesworth back at ground control must have been gobsmacked: "The *Eagle* has wings," "The *Eagle* has landed," and now this. From *Neil.*

A last word, though, about the "a" before moving on. Neil later said "We'll never know" about what he had earlier said, but we at LIFE say he said "a man," and we cite the good reporting of *NBC News* space analyst James Oberg. A dedicated mythbuster when it comes to space program lore and legend—he even knew the Mr. Gorsky gag was Buddy Hackett's—Oberg recalled upon Armstrong's death, "I personally heard the broadcast live while I was a 'NASA trainee' at Northwestern University's Technological Institute in Evanston, Illinois, and when I immediately repeated the line for a colleague, I distinctly recall saying it as I had interpreted it: 'That's one small step for a man . . .'"

Good enough for us.

THE EAGLE'S TWO-MAN crew went on to perform each of its duties on the lunar surface without surprise or malfunction. Toddling along in space suits built to withstand temperatures from 250 degrees Fahrenheit below zero to 250 degrees above, as well as any meteorites that might be zipping about, Armstrong and Aldrin collected 50 pounds of rock and soil samples. They set up a seismometer to measure earthquake-like activity, a reflector to pick up moon-bound laser beams sent from earth to measure the exact distance (turns out, 239,000 miles) and a sheet of foil to trap gases for study back home. Armstrong and Aldrin spent two hours and 31 minutes walking on the moon before they clambered back into the *Eagle,* which lifted off serenely and joined the mother ship.

For the next three days, they and Collins cruised "right down U.S. 1," as Armstrong put it. "Very smooth, very quiet ride." On July 24, after crashing through earth's atmosphere at some 25,000 miles an hour, generating temperatures nearing 4,000 degrees Fahrenheit, they splashed down in the Pacific, 15 miles from the recovery ship. They were entirely surrounded by a world in awe.

That wasn't the end of it, of course. As mentioned briefly in our last chapter, the saga of *Apollo 13,* marooned in space before being dramatically reeled back to earth, would follow in 1970. In 1972 the Apollo program would close up shop; the final tab would come to $20.4 billion. NASA's recyclable-craft shuttle program would be marked by a string of glorious successes—a cosmic fix-it operation for broken satellites, a taxi to the space stations—and the two horrifying explosions before it, too, went out of business in 2011. The Hubble telescope would probe deep space; Pathfinder and its successors would find the surface of Mars.

But all these years later—these 43 years later, as we look at the life of Neil Armstrong—one space adventure still seems magical.

Do you believe we put a man on the moon?

CAUSE FOR CONCERN
The problem with a deadline, such as the one imposed eight years earlier by President Kennedy and now the one being adhered to by NASA's bosses, who are hoping for a space-program comeback (and know the Soviets are hoping for the same), is that when things go wrong, you fix them on the fly. The sequence below shows Armstrong's bad luck—and also, great luck—in the episode with the lunar landing research vehicle, right, in 1968. His deft ejection at the last second might have helped secure him important assignments on the big mission, but the memory of the smoldering vehicle on the ground lingers for all who witnessed it. In the event, of course, Armstrong would have to wrestle with the real lunar landing module in the very real moment, and, with less than a minute's worth of fuel remaining, would come through magnificently.

To the Moon and Back

NO STONE UNTURNED

Houston, Iceland, Cape Kennedy in Florida . . . all manner of simulators, indoor moonscapes that might as well be used to shoot a sci-fi flick . . . exercise, medical exams and training, training, training. It must have seemed at the time to Armstrong (in all three pictures) and Aldrin (below, right, and at left in the photograph opposite) that things would be so well-rehearsed they might be boring by the time they were descending in the lunar module, or kicking up real lunar dust instead of good old American sand. "I have confidence in the equipment, the planning, the training," Armstrong told LIFE. "I suspect that on a risk-gain ratio, this project would compare very, very favorably with those to which I've been accustomed in the past 20 years."

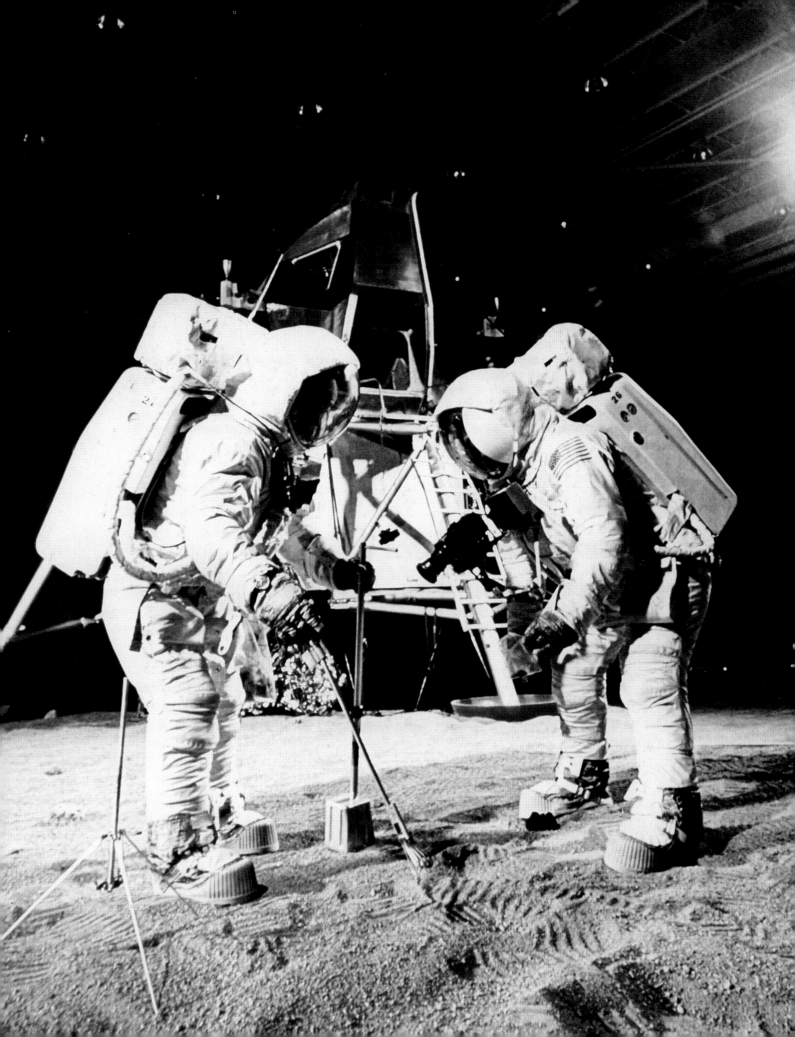

BEACH PARTY

All back roads and front yards in and around Cape Kennedy were jammed with all manner of vehicle and all kinds of people even before dawn. Across the way on Pad 39A the *Apollo-Saturn* stack loomed awesomely as the sun came up. People imagined it was itching to go, couldn't wait to get to the moon. The astronauts were, of course, removed from this scene. They awoke, cleaned up, breakfasted together, got dressed (or suited) and were escorted to their vehicle. They took their places in the capsule. Everything seemed to be working. Everything *was* working. The countdown was counted down, and then . . . The earth began to shake, and the rumble was heard, and the huge rocket began to rise.

LEVITATION

LIFE's Ralph Morse hung a remote camera to capture a rocket launch in a way no one ever had. LIFE also ran, in August 1969, 25,000 words by Norman Mailer under the title "A Fire on the Moon." Mailer wrote that "the liftoff itself seemed to partake more of a miracle than a mechanical phenomenon, as if all of huge Saturn itself had begun to levitate, and was then pursued by flames. No, it was more dramatic than that. For the flames were enormous. No one could be prepared for that . . . Flames flew in cataract against the cusp of the flame shield, and then sluiced along the paved ground down two opposite channels in the concrete, two underground rivers of flame which poured into the air on either side a hundred feet away, then flew a hundred feet further. Two mighty torches of flame like the wings of a yellow bird of fire flew over a field, covering a field with brilliant yellow bloomings of flame; and in the midst of it, white as a ghost, white as the white of Melville's Moby Dick, white as the shrine of the Madonna in half the churches of the world, this slim angelic mysterious ship rose without sound out of its incarnation of flame and began to ascend slowly into the sky, slow as Melville's Leviathan might swim, slowly as we might swim upward in a dream looking for air. And still no sound." After the third part of Mailer's LIFE series concluded in February 1970, the never modest author wrote to the ever modest Neil Armstrong: "I've worked as assiduously as any writer I know to portray the space program in its largest, not its smallest, dimension."

RALPH MORSE

BACK HOME

It could have been that Jan Armstrong, watching the television coverage with her son Ricky, thought back to the anguish she had felt during the *Gemini 8* mission. There would be tense moments during this one, too, although how dicey the landing module descent became was, initially, known only by those in Houston, and those aboard *Eagle*. ("I am told that my heartbeat increased noticeably during the lunar descent," Armstrong wrote later in LIFE, "but I would really be disturbed with myself if it hadn't.") On this page, above right: Once the mission has been safely launched and his dad is well on his way to the moon, Ricky puts up the flag. Left: Friends and neighbors gather at the house of Stephen Armstrong (Neil's father) in Wapakoneta to watch the moon walk. There had never been a television broadcast like this before, and in 43 years, there hasn't been one like it since.

To the Moon and Back

A LAND OF STRANGE LIGHTS

Armstrong wrote in LIFE: "It took us somewhat longer to emerge from *Eagle* than we had anticipated but the delay was not, as my wife and perhaps some others have half-jokingly suggested, to give me time to think about what to say when I actually stepped out onto the moon. I had thought about that a little before the flight, mainly because so many people had made such a big point of it. I had also thought about it a little on the way to the moon, but not much. It wasn't until after landing that I made up my mind what to say: 'That's one small step for a man, one giant leap for mankind.' Beyond those words I don't recall any particular emotion or feeling other than a little caution, a desire to be sure it was safe to put my weight on that surface outside *Eagle*'s

footpad." The boot above is Armstrong's. He continued in LIFE: "From inside *Eagle* the sky was black, but it looked like daylight out on the surface and the surface looked tan. There is a very peculiar lighting effect on the lunar surface, which seems to make the colors change. I don't understand this completely. If you look down-sun, down along your own shadow, or into sun, the moon is tan. If you look cross-sun it is darker, and if you look straight down at the surface, particularly in the shadows, it looks very, very dark. When you pick up material in your hands it is also dark, gray or black. The material is of a generally fine texture, almost like flour, but some coarser particles are like sand. Then there are, of course, scattered rocks and rock chips of all sizes."

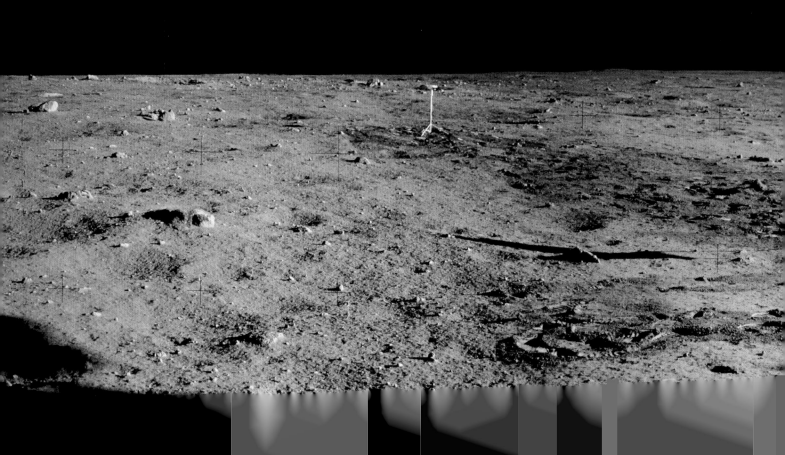

To the Moon and Back

AN INTERESTING PLACE TO BE

Most of the photos from the moon are of Aldrin, taken by Armstrong (with Armstrong sometimes reflected in Aldrin's visor). Most, but not all. On the opposite page, top, Armstrong and Aldrin fix a flag in place (it has a rod horizontal at the top so it will appear unfurled; there is no breeze to blow it). That picture was taken by a camera on the lunar module. In the panorama below, taken by Aldrin, Armstrong works near the module. At left, he smiles once back aboard the lunar module, mission halfway accomplished. (Now, let's get safely home!) Armstrong wrote in LIFE that in ascending from the lunar surface, *Eagle* "gave us not only a very pleasant ride but it also afforded us a beautiful, fleeting, final view of Tranquility Base as we lifted up and away from it." He later spoke in a similar way during an interview on *60 Minutes:* "It's a brilliant surface in that sunlight. The horizon seems quite close to you because the curvature is so much more pronounced than here on earth. It's an interesting place to be. I recommend it."

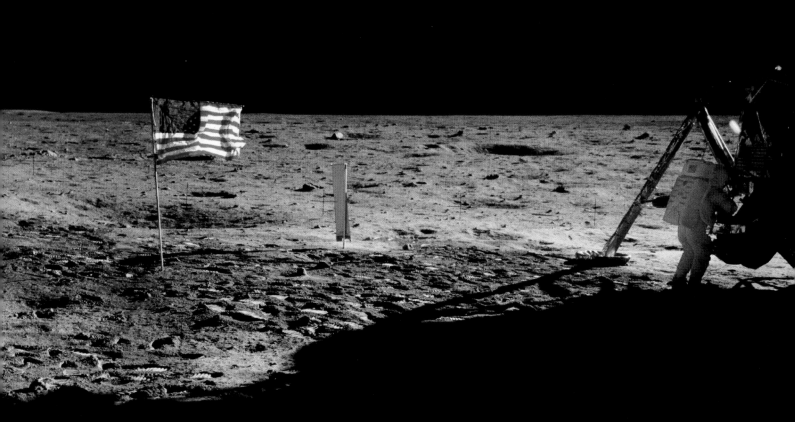

To the Moon and Back

AS THE ASTRONAUTS WALKED, THE EARTH STOOD STILL

The Pope paused to watch at his summer villa in Italy (below), and so did the gamblers at the gaming tables in Vegas, Monte Carlo and everywhere else (right). A crowd gathered around a TV set at a duty free shop at Mascot Airport, way down under in Sydney, Australia (bottom), and 10,000 cheered as they watched giant screens up at New York City's Central Park (below, right). Five hundred million people stopped to watch Armstrong and Aldrin's great adventure. One who couldn't, interestingly enough, was *Apollo 11*'s Michael Collins, who stayed in steady orbit around the moon while his partners gamboled below. "I never caught a glimpse of *Eagle* on the surface of the moon," he wrote in LIFE, "but I could sometimes hear them . . . People keep asking me if I was lonely up there in *Columbia* while Neil and Buzz were on the moon. I wasn't. I've been flying airplanes by myself for about 17 years, and the idea of being in a flying vehicle alone was in no way alarming. In fact, sometimes I prefer to be by myself."

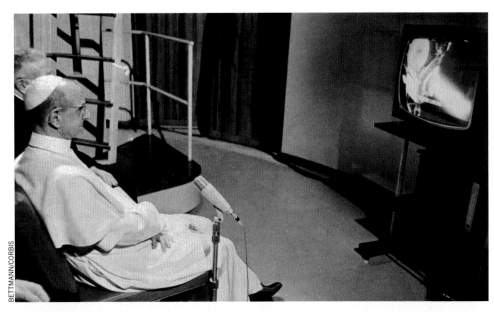

To the Moon and Back

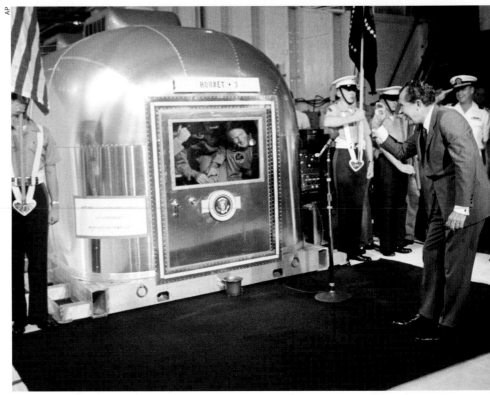

FROM ONE CAPSULE TO ANOTHER
After being plucked from the sea and put in special suits (above), the astronauts are quarantined in a special van aboard the USS *Hornet,* where President Richard Nixon has been waiting to greet them, as best he can (right). "The quarantine period was a little burdensome," Michael Collins complained in LIFE. "We were really glad to hear that the mice didn't pick up any moon bugs." If Nixon was over the moon, the reaction in Moscow was something other (opposite), which is not what Armstrong would have wished. Of all the things he and Aldrin left behind on the moon, he wrote in LIFE, "We were particularly pleased to deposit the patch of Apollo 1 in memory of our friends and fellow astronauts Gus Grissom, Ed White and Roger Chaffee, and the medals that were struck in commemoration of [Soviet cosmonauts Yuri] Gagarin and [Vladimir] Komarov. I believe that those gentlemen and their associates share our own dreams and hopes for a better world . . . I was encouraged in this belief by a telegram of congratulations which was waiting for us in the Lunar Receiving Laboratory when we returned. It began 'Dear Colleagues,' and it was signed by all the cosmonauts who have flown." Not exactly what JFK had in mind when he launched this plan back in '61.

BETTMANN/CORBIS

AP

THE RITUAL

Many Yankees and Giants and even a few Mets and Jets have traveled the Big Apple's byways (above), and several other astronauts have too, but the confetti parade in New York City that greeted the astronauts was, like everything else surrounding the *Apollo 11* flight, truly memorable. This celebration came early in a brutal globe-trotting road show, and few people knew at the time just how little the star of the show, Armstrong, at right with Jan during another parade back in the good old hometown, Wapakoneta, Ohio, cared for this kind of adulation. It made him question why he did what he did, who was responsible for his success. As Jan observed, it tore him up inside. It might have been that he was more comfortable on the moon.

The Quiet Hero

KNOWING WHAT WE know now about the kind of man Neil Armstrong was—the more complete picture of the man who cherished privacy, rather than just "the first man on the moon"—his victory lap in 1969 must have been a certain kind of hell: fetes and speeches in Washington, a tour of 28 cities in 25 countries in barely more than a month, audiences with the queen of England and the Spanish dictator Francisco Franco among many, many others. "It's not that I feel uncomfortable," he said politely at a NASA press conference in 1970 when asked about this new hero status. "It's just that I find there's inadequate time to do all the things I'd like to be doing."

He would act, and pretty quickly, to solve the problem. As Kathy Sawyer wrote in her *Washington Post Magazine* profile on the occasion of *Apollo 11*'s 30th anniversary: "Armstrong coped with the new demands of fame by means of his finely calibrated personal code that, while sometimes mystifying to outsiders, made perfect sense to him. He would do his duty as he saw it, he would be conscientious, he might participate in causes, events and moneymaking ventures that he deemed worthwhile. But he would try to avoid anything that focused on him, his personal life or his celebrity. There have been complaints that the first man on the moon should have been a more aggressive and accessible ambassador for the space program. But crewmate Collins, for one, has supported Armstrong's approach, arguing that Armstrong, more than anyone, seemed to understand his unique role and the need to ration himself.

"'I think he saw the results of being an idol when he researched Lindbergh's experience,' says Jim Lovell, hero of *Apollo 13*, who trained with Armstrong and has kept in touch with him over the years. 'He didn't want to have his life change. He decided to be very reclusive, but that's also his nature.'"

Lovell understood Armstrong well, but

GIVING SUPPORT, PERHAPS HOPE

Not long after he has returned from the moon in 1969, Neil Armstrong is happy (in this instance) to be shaking hands. He is in Long Binh, Vietnam, as part of, yes, Bob Hope's annual Christmastime tour for the troops. More than 20,000 servicemen and -women give naval veteran Armstrong a standing ovation this day, and one has come prepared with a space atlas, which Armstrong autographs— as always, painstakingly, with that strong signature seen on the yearbook page pictured at the beginning of our book.

The Quiet Hero

others at NASA surely didn't. They paid him to run a department in Washington and gave him a fancy title—deputy associate NASA administrator for aeronautics—and a corner office with a view of the Capitol. But what they really wanted was a Cartier-quality glad-hander near the seat of power, someone they could trot out as needed: the very definition of anathema to Armstrong. In 1971 he went back home to Ohio, bought a dairy farm in Lebanon and took a teaching post at the University of Cincinnati a half hour away. A phalanx of reporters and cameramen showed up for his first day of work, and once he had ushered his students into the classroom, he slammed the door shut.

During his decade at the university he worked hard as a teacher, tried to do what was expected of him, tried to avoid being made into more of a public figure than he already was. He gave precious few interviews. His reputation became something of a bizarre cross between Gary Cooper and J.D. Salinger. That was a product of what Sawyer called "his finely calibrated personal code," in force day after day.

Until, finally, his famous reticence established a serviceable buffer, and he could say yes again. He started to give more speeches and accept certain honors. (He did both in October 2007 when his alma mater, Purdue, dedicated the new Neil Armstrong Hall of Engineering.) He joined various corporate boards of directors, and developed a successful sideline career in business. He testified before congressional committees on behalf of NASA projects and, in a recent instance, he came out strongly against President Obama's cancellation of a program to further lunar exploration. Rather astonishingly, in the past few years he took to the road with fellow Apollo veterans Gene Cernan and Jim Lovell, visiting Boy Scout troops in New York City and military troops overseas.

He watched his children grow. He and Jan divorced in 1994. (She lives today in Utah.) In 1999 he married Carol Knight, a widow at the time, and they lived in the Cincinnati suburb of Indian Hill.

Some things that were true when he was six and flew in that barnstorming Ford Trimotor—the *Tin Goose*—with his father remained true when he died at age 82. He was still shy, still withdrawn. He was still inquisitive. He was still dutiful. He still loved flight.

His family—deep down, as poetic as he was generally and as he proved himself to be on one special day—acknowledged all

---

A BUCKEYE AT HEART
Ohio welcomed him home. After that brief stint as an antsy Washingtonian, Armstrong moved, along with his family, back to the Buckeye State in 1971, to the farm in Lebanon seen at right. Meantime up the highway in Wapakoneta, tribute-building had already begun for the native son; it has never ceased. For years now, people have driven through the intersection of Saturn and Apollo, and since 1972 have visited the Armstrong Air & Space Museum.

of this when he passed:

"He remained an advocate of aviation and exploration throughout his life and never lost his boyhood wonder of these pursuits . . . As much as Neil cherished his privacy, he always appreciated the expressions of good will from people around the world and from all walks of life. While we mourn the loss of a very good man, we also celebrate his remarkable life and hope that it serves as an example to young people

around the world to work hard to make their dreams come true, to be willing to explore and push the limits, and to selflessly serve a cause greater than themselves."

Read those words again.

Neil Armstrong, once upon a time having reached an amazing place and needing to say something about it, had all of that in mind when he declared that, perhaps, one small step for a man could, somehow, become a giant leap for mankind.

A MAN OF MANY JOBS

Opposite, top: Professor Armstrong, on his first day of work at the University of Cincinnati in 1971, is about to shut the classroom door—firmly. Above: In the eternal role he could not shake simply by retiring from NASA, this role being "the first man on the moon," Armstrong visits the Dryden Flight Research Center in Edwards, California, where he used to fly X-15s, and climbs into the cockpit of an SR-71. Opposite, bottom: On February 27, 1986, he is in the front row at left while serving as the vice chairman of the Rogers Commission (that's former U.S. Secretary of State William P. Rogers beside him), a presidential committee probing the disaster that claimed the space shuttle *Challenger* and all aboard. On the other side of Rogers is former astronaut Sally Ride, who was the first American woman in space and who, by chance, predeceased Armstrong by only a matter of weeks in the summer of 2012. Back in '86, their commission's report found problems in NASA's communications and procedures as well as design flaws in the vehicle, and said the tragedy was "an accident rooted in history."

WHEN HE CHOSE TO ATTEND

Armstrong hardly turned into a hermit in his astronaut afterlife, and there's no question but that he was proud of what he—and his many associates—had accomplished. If the particular event was important or the cause was worthy, Armstrong would show. Opposite, from top: On October 1, 1978, at the Kennedy Space Center in Florida, he receives the first Congressional Space Medal of Honor from President Jimmy Carter, assisted by, at left, Captain Robert Peterson; on July 20, 1994, at the White House, Armstrong, Aldrin and Collins shake President Bill Clinton's hand at a 25th anniversary celebration of the moon landing; and, on the same date in 2009, the three men, this time with Aldrin on the left, Collins in the center and then Armstrong, celebrate the 40th anniversary of the same event in the same building, President Barack Obama presiding in the Oval Office, where George Washington looks on—amazed at all that has transpired. Above: On November 16, 2011, the three moon men (pictured are Aldrin, left, and Armstrong) plus John Glenn are awarded Congressional Gold Medals in the Capitol Rotunda.

"A Still More Glorious Dawn Awaits" —CARL SAGAN

NEIL ARMSTRONG HALL OF ENGINEERING

Neil Armstrong
BS Aeronautical Engineering 1955
Honorary Doctorate 1970

AND A LAST TRIBUTE

He received, in his lifetime, all the plaques and citations, medals and tributes anyone could ever want—and many more than he wanted, certainly. Two in the Heartland that surely meant something to him (this is clear simply because he allowed them to happen) were the Air & Space Museum that was built in his hometown of Wapakoneta, Ohio, and, seen here, the school housed within the Neil Armstrong Hall of Engineering on the campus of his alma mater, Purdue University, in West Lafayette, Indiana. The flowers and candles were left in the last week of August 2012 when it was learned that Armstrong had passed away, and there were many more just like them scattered throughout Wapakoneta, where on Wednesday, August 29, there was a moving memorial service held. As for this little girl, her name is Kirsten South. She may study here one day . . . Or she may be a pilot . . . Or an astronaut. She may go to the moon. As Neil Armstrong proved: Nothing's impossible.

MOON MEN ROAMING THE EARTH
Neil Armstrong and Gene Cernan had walked on the lunar surface, and Jim Lovell, veteran of four NASA space missions, was the only man ever to have traveled to the moon twice and never touched down (the famous fate of *Apollo 13* is known to all). Beginning in 2010, these former colleagues and lifelong mates engaged in a series of morale-boosting tours, one of which is remembered in Jeffrey Kluger's essay. Front row, from left, Cernan, Lovell and Armstrong pose with the troops at a U.S. military base in Kuwait.

"The Armstrong I Saw"

By Jeffrey Kluger

I **DON'T KNOW** how many times an airplane tried to kill Neil Armstrong, but with 78 combat missions in Korea and a test-piloting career that saw him at the stick of 900 flights, it's a fair bet he got used to the experience. I do know what was probably the last time he faced that kind of danger, because I was along for the ride—and so were a couple of hundred other people.

It was in March 2010 and we were part of a large group returning from a six-country morale tour of military bases in the Middle East—a tour that also headlined Apollo veterans Gene Cernan and Jim Lovell. The trip included a flight over Iraqi airspace and down the Persian Gulf as well as a tailhook landing on the deck of a carrier and a catapult launch off the next day. The last leg of the journey, at least, would be routine: an overseas flight from Ramstein Air Base in Germany to Kennedy Airport in New York. Or at least it was supposed to be routine.

The weather forecast for much of the East Coast was brutal that day, and the reality was living up to those predictions, with lashing rains and powerful, almost lateral winds. The approach to the airport was murderous, with the plane lurching, the flight attendants urging calm. You could see the wings, to the extent they were visible through the soup outside the window, slightly but literally flapping. I had gotten lucky and snagged a seat in business class, but I could hear screaming coming from the rear of the plane, where the buffeting was worst. The astronauts were up in first.

As we made our final approach—with the tarmac in sight just below the wheels—there was a sudden gust of wind, the plane slewed to the side, and the pilot pulled suddenly up and broke off the landing. He climbed, circled back around and tried again. The result was the same—perhaps even worse. This time when he climbed, he gave up and headed off for a safer landing in Boston.

The scene in the cabin was what you'd expect, with loose belongings littering the aisles, air-sickness bags everywhere and the few children onboard wailing inconsolably. As soon as I could, I unbuckled and walked shakily up to the front. The first person I saw was Armstrong. He was sitting in his seat, his newspaper still open in front of him, serenely working on a Sudoku puzzle.

In some ways, one would expect nothing else. You don't pilot war planes, experimental jets and spacecraft without knowing how to retain your composure when things get dicey. But it was the depth of Armstrong's composure that defined the man. Lovell and Cernan were untroubled too, but like all of the other passengers in what had been a marginally less storm-tossed first class, they were talking and even laughing with that high and happy release of tension that always follows danger. Armstrong was doing what Armstrong did best: remaining contained, controlled and wholly, utterly private.

That may have been part of his nature, but his experiences drew it out of him, too. Unlike most historical giants, Armstrong had his place in history bookmarked for him long before he got there. "The first man on the moon" had been a cultural construct for centuries—someone who would surely become real one day, but only when humanity had developed the improbable technology to make it so. No child ever dreamed that they would one day invent the light bulb or explain relativity—though when Edison and Einstein did so, they certainly joined the gallery of history's greats. But every child—particularly every boy—at some point angled for the gig Armstrong landed.

And yet when it finally happened, something had subtly changed. "ARMSTRONG TO BE FIRST MAN ON MOON," ran the headlines when NASA released its flight manifest for 1968 and 1969. Armstrong had worked hard and competed ferociously to land that spot, but ultimately, it was others who tapped him for fame.

That, of course, was the only way it could have been. The likes of Charles Lindbergh and the Wright brothers

"The Armstrong I Saw"

succeeded more or less under their own steam and on their own dime. Armstrong's mission required a global web of industrial, political and technological support—an invisible army of some 400,000 people. Astronauts like Armstrong simply walked point, which was fine; someone had to. But Armstrong also got to—*had to*—wave in parades while the others went unthanked. That never ceased to trouble him. "He feels guilty that he got all the acclaim for an effort of tens of thousands of people," said his first wife, Janet, in James R. Hansen's authorized biography. "He's certainly led an interesting life. But he took it too seriously to heart."

During an earlier leg of our Middle East tour, I got a glimpse of Armstrong's deep discomfort with laurels he perceived were unearned. I asked him if I might get in touch with him after we got home to conduct an interview or two for a piece I was considering writing—one that would include some details about his past. "I confess that I haven't yet read your book," I said, referring to Hansen's *First Man: The Life of Neil A. Armstrong.*

"I don't have a book," Armstrong said. Snapped, actually. That was true; he hadn't written a book, and to say otherwise seemed to offend him on two levels. It would give him credit—again!—for work someone else had done, and it would imply that he was the kind of person who would write an autobiography in the first place. Both prospects were anathema to Armstrong—and he paid a personal price for that. When Hansen commented to Janet that Neil had won universal respect for never cashing in on his fame, her response was succinct: "Yes, but look what it's done to him inside."

There was a certain monastic power to Armstrong's famous reserve, on display many times in the previous pages of this book. It seemed that every time his moon-walking crewmate Buzz Aldrin appeared on a sitcom or an autograph show or at this or that glittering dinner, Armstrong would retreat a bit more, as if there were a fixed level of quiet grace his crew would be forever required to maintain. Whenever Aldrin tacked one way, he would have to tack the other.

That made it hard, at first, for me to understand why this most private of men had gone along on a meet-and-greet Middle East tour at all—and why on two occasions afterward, he went back to Iraq and Afghanistan with Lovell and Cernan for more of the same glad-handing and public speaking he seemed to abhor. But for Armstrong at least, this may have been a way of paying down a perceived debt: tithing his time to balance a moral ledger that only he was keeping. And yet there were other, subtler debts he probably reckoned he would never quite be able to pay.

During our visit to Ramstein, a very young boy—the son of a serviceman—was introduced to Gene Cernan. The father explained to his son that the man in front of him had walked on the moon—and had even lived there for a few days! The boy looked at Cernan with delight.

"Are you Neil Armstrong?" he asked.

Cernan, to his credit, laughed. "No, son. I'm Gene Cernan. But I went to the moon too."

The little boy turned to his father, puffed with indignation. "You said Neil Armstrong!" he protested.

Cernan, of course, was used to this. Lovell was used to it, Aldrin was used to it—the whole grand group of two dozen men who sailed moonward 40 years ago, did their surveying, took their pictures, collected their rocks, planted their flags and came safely and improbably home, were used to it. And Armstrong, surely, was more used to it than all of them. Others may have been able to carry that singular burden more easily and lightly than he did. But no one could have carried it with greater, more resolute strength. Armstrong had everything in his life a man could have wished for, except, perhaps, a sense of deep peace. If there is any consolation to be had from his passing, it is the hope that he has that now.

JEFFREY KLUGER is the senior science writer at *Time* magazine and coauthor, with Jim Lovell, of *Lost Moon: The Perilous Voyage of Apollo 13,* and other books.

GIVING AND RECEIVING
Armstrong certainly wouldn't have agreed in 2010 to participate in the tours if he hadn't thought them worthwhile for the men and women out there, or if he'd thought he was being exploited. In the event, the astronauts were enriched, too. Above: Armstrong is given the Naval Astronaut Wings by the captain of the USS *Dwight D. Eisenhower* as Cernan looks on. Opposite, top, from left: Armstrong, Lovell and Cernan greet a soldier onboard. Bottom: Armstrong on the ship's deck.

Shining On

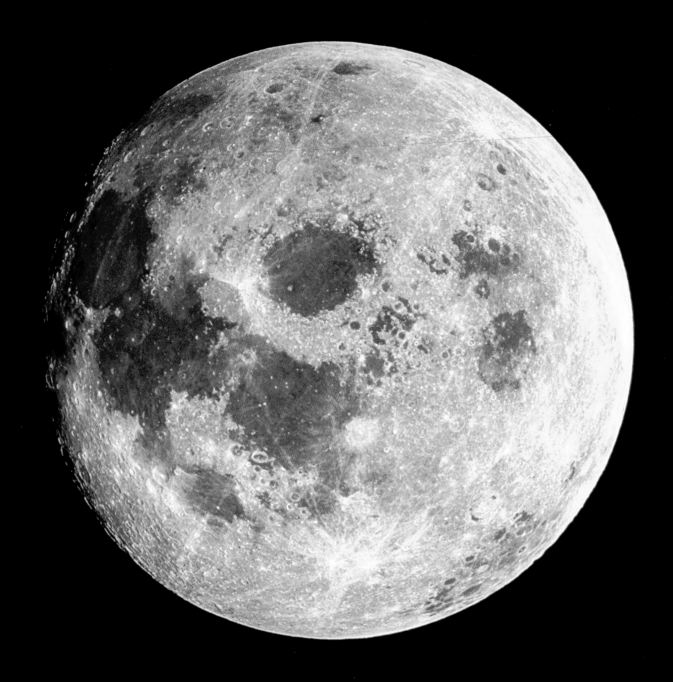

This picture was taken in 1969 from *Apollo 11*, commanded by Neil Armstrong, who would forevermore be linked in the public's mind with the moon. He was okay with this, and his family shared his good humor about the situation. When their patriarch passed away in August of 2012, they thanked the millions who had sent along expressions of pride and sympathy. The family suggested that proper gestures might include a rededication to the virtues of service, accomplishment and modesty Armstrong so embodied, and then they added, "And the next time you walk outside on a clear night and see the moon smiling down at you, think of Neil Armstrong and give him a wink."